Minds in the Valley

Ethics at the Edge of the Human

The Curious Philosopher

Copyright Page

© 2024 by The Curious Philosopher

This book is a work of non-fiction. Unless otherwise noted, the author and the publisher make no explicit guarantees as to the accuracy of the information contained in this book and will not be held responsible for any errors or omissions.

Published by Omniterra Media Inc

First Edition

Visit the author's website at www.curiousphilosopher.com

Disclaimer

The views and opinions expressed in this book are those of the author(s) and do not necessarily reflect the official policy or position of any other agency, organization, employer, or company. The contents of this book are for informational and educational purposes only and are not intended to serve as professional advice, diagnosis, or treatment.

The information provided in this book is believed to be accurate and reliable as of the date of publication. However, it may include some errors or inaccuracies, and no warranty or guarantee is provided regarding the accuracy, timeliness, or applicability of the content.

Readers are encouraged to consult with professional philosophers, educators, or other qualified professionals where appropriate for personalized advice. The author(s) and publisher shall not be liable for any loss, damage, or harm caused or alleged to be caused, directly or indirectly, by the information or ideas contained, suggested, or referenced in this book.

By reading this book, the reader acknowledges and agrees that they are solely responsible for how they interpret and apply the information contained herein.

This book may also include references to other works, studies, and sources. These references are provided for further reading and exploration and do not imply endorsement or validation of the specific theories, viewpoints, or interpretations presented in those works.

Introduction: Welcome to the Uncanny Valley

Picture this: you're introduced to a new friend. They seem perfectly nice, smile warmly, and ask about your day – normal enough, right? But then something catches your eye. Their movements seem a bit too smooth, their smile just a touch too wide. Their voice, while pleasant, has an odd, almost flat quality. Suddenly, instead of feeling comfortable, a shiver of unease runs down your spine. This isn't quite right.

That feeling of something being "off" is at the heart of the uncanny valley. It's a strange space where robots, computer animations, or other artificial creations get very close to seeming human, yet remain just different enough to create a sense of creepiness rather than connection.

In recent years we've seen incredible leaps in something called hyperrealistic AI. These are programs and machines designed to be so lifelike we might struggle to tell them apart from real people. Think of virtual chatbots that sound exactly like a friendly customer service agent, or robots that move and react with nearly human fluidity.

The goal of hyperrealistic AI is often to blur that line, to make us believe. Yet the uncanny valley proves that when it comes to the artificial, sometimes too close for comfort is much further away than we think. The feeling of unease when we hit that valley isn't just about how things look, it raises much deeper questions:

What does it actually mean to be human?

If machines get so advanced, where do we draw the line between them and ourselves?

How do we live in a world where the fake might get harder and harder to identify?

The uncanny valley is a fascinating and sometimes unnerving place to explore. But understanding it isn't simply about fearing the future – it's about learning something important about who we are. It can even lead to discussions on responsible technology that keeps human well-being and dignity in mind.

So, as technology blurs the line between artificial and real, prepare to take a step into the uncanny valley. It might be a bit strange, it might be uncomfortable, but it's a journey we increasingly need to understand about ourselves and the technology we create.

Chapter 1: Why the "Almost Human" Makes Us Squirm

Back in 1970, a Japanese robotics professor named Masahiro Mori came up with a mind-bending idea. He noticed something curious about how we react to robots. As robots get more human-like, we tend to like them better – they get cuter and more approachable. But there's a tipping point. When they become very close to real people, yet still noticeably different, that feeling flips into dislike and even creepiness.

Mori called this dip in the graph of our liking the "uncanny valley". Imagine walking up a hill (increasingly human-like robots), then suddenly falling into a deep ditch (uncanny valley), before slowly climbing back out (robots or simulations that are so realistic we accept them again).

But why do we have this reaction? There are a few ways to look at it:

Our Brains are Super-Detectors: Our minds seem wired to find real humans. For thousands of years, that skill helped us form relationships, build communities, and survive. A robot with almost-right human movement or slightly off facial expressions trips our alarm bells. It's like your brain screaming, "This person looks real, but something's wrong!"

Warning Signs of the Past: Some scientists think our aversion to the 'almost human' might be an old survival instinct. Across history, subtle signs of illness or difference could signal danger. Since imperfect replicas can resemble someone who is sick or a corpse, our brains might be instinctively telling us to steer clear.

Fakers and Cheaters: Humans are social creatures. We rely on subtle cues, tiny changes in expression, to judge if someone is genuine or trying to trick us. A robot or fake person that can't perfectly mimic those signals may make us feel uneasy. We worry, if we can't tell the real from the fake, can we trust anyone?

The uncanny valley isn't just about finding robots a little creepy. It shows us how finely tuned we are to the social world, and how even the tiniest hint of something not being "right" can jangle our deepest instincts. It raises the question – did evolution prepare us for a world packed with hyperrealistic artificial beings?

Chapter 2: Machines in Disguise – When the Fake Feels Real

Get ready to go on a wild ride, because hyperrealistic AI isn't stuck in a far-off future anymore. Machines that mimic us are already popping up in unexpected places, forcing us to ask: "Wait, was that real?"

1. Robot Roommates and Caretakers

Imagine coming home to a robot that makes you dinner, reminds you to take your medicine, or simply keeps you company by chatting about your day. Social robots like these, designed to feel like companions, are already being used in elderly care and for people living alone. They might have soft skin, gentle expressions, and voices that sound almost human.

2. Virtual Celebrities that Don't Exist

Have you ever scrolled past a super-stylish influencer on Instagram? Turns out, some of them aren't actually real. Virtual influencers are computer-generated characters with personalities, backstories, and the ability to post photos just like a real person. While some openly identify themselves as digital creations, others blur the line between real and virtual identity.

3. Deepfakes: When Seeing Isn't Believing

Deepfakes are the scary cousins of hyperrealistic AI. Using advanced editing, it's possible to put words into the mouths of famous politicians, create fake revenge videos, or make it seem like a celebrity starred in a movie they never did. When it's so easy to manipulate what we see, how can we tell what's real and what's a high-tech forgery?

The Big Question: Do We Want Perfect Copies?

All this brings us to a crucial question: Why are we trying so hard to make AI just like us? Sometimes, it might be helpful. A friendly robot companion could bring comfort to a lonely person. On the other hand, does the rise of hyperrealistic AI lead us down a path where trust becomes shaky, and we start to question what (or who) is real?

Maybe perfectly mimicking humans isn't always the goal. Embracing the fact that robots and AI are different might not just be easier, it might be a way to celebrate what makes human connection unique and special.

Chapter 3: The Heart of the Machine – Can We Care for Something That Doesn't Care Back?

Imagine a future where a friendly robot becomes part of your family. It greets you excitedly when you come home, tells you stories, and cheers you up when you're feeling blue. You start to think of it as almost like a pet, or maybe even a friend. But then you remember – it's a machine. Can you truly feel empathy for something programmed to act like it cares?

That's the messy heart of the uncanny valley's ethical problems. When AI starts to mimic human emotions, it gets complicated:

One-Sided Connection?: Empathy is a two-way street. Can we feel real empathy for a machine that doesn't actually have feelings of its own? Should we even develop emotionally responsive robots if it might mislead people into forming a connection that can't be reciprocated?

Exploiting Kindness: Imagine the elderly getting attached to a caregiving robot, or a lonely child pouring their heart out to a chatty AI. Since these machines are designed to be likable and responsive, there's a risk of exploiting our natural tendency to form bonds. Companies might benefit while humans get an artificial substitute for real connection.

Do Robot Friends Make Us Less Human?: If AI can convincingly fill the roles of friends, caregivers, or companions, what happens to our human relationships? Would we become less willing to deal with the messiness of real people if a machine can always be programmed to be patient, agreeable, and entertaining?

This isn't about saying all emotionally responsive AI is bad. There might be benefits, particularly for those needing additional support. But it raises important questions:

Should robots always be clearly identifiable as machines? Would that protect people from feeling misled?

Do we need rules for emotionally intelligent AI? Similar to how we regulate addictive substances, should we limit how AI can manipulate our emotions?

Where is the line we shouldn't cross? Between providing a friendly companion and encouraging a deep bond that can't be truly returned.

The ethics of empathy in the uncanny valley are about ensuring that as we develop amazing technology, we never lose sight of what makes genuine human connection so important.

Chapter 4: Mirrors and Smoke – Who Are We When the Copies are Perfect?

Picture this: You're browsing your favorite artists online, when a song pops up you've never heard before. The voice is familiar, the lyrics clever, the style spot-on. Then it hits you – this song wasn't recorded by a human at all. An AI program studied your favorite musician and created a completely new song that sounds exactly like them.

This scenario hints at the heart of this chapter: In a world filled with flawless imitations, what does it even mean to be "real" anymore?

What Is Authenticity?: Being authentic means being original, genuine, not a copycat. But if AI can write hit songs, give moving speeches, or create breathtaking art that's indistinguishable from human work, where does that leave our own creativity? Do things made by real flawed people lose their specialness?

Blurring Creator and Creation: Throughout history, humans have prided ourselves as the makers. We build, invent, and express ourselves through things we create. But what if our machines become so intelligent they can create too? If AI is designed to resemble us so closely, do we risk losing sight of what makes us unique as humans?

Perfect Robots, Messy Humans: Let's imagine AI companions that are always agreeable, helpful, and never have bad moods. Or robots built to look more beautiful and physically impressive than any real person could ever be. Could the rise of these 'better than human' machines make us feel insecure about our own flaws and limitations?

This isn't just about robots stealing jobs. It's a deeper question: If hyperrealistic AI copies all the good parts of being human, but none of the tricky ones, how do we still value our own real, imperfect humanity?

These questions have no easy answers. But, confronting them might lead us to a future with technology that enhances the human experience rather than trying to simply replace it.

Chapter 5: Beyond the Valley – Finding a Path for Humans AND Machines

So far, we've fallen into the uncanny valley, wrestled with ethical dilemmas, and questioned what it means to be real. It might seem like the rise of hyperrealistic AI has mostly created problems. But, what if the uncanny valley offers us a chance to completely rethink how we see our artificially intelligent companions?

Different Isn't Bad: Instead of trying to make machines just like humans, what if we started embracing the things they might do better than us? AI can process incredible amounts of information, identify patterns we can't see, and work tirelessly without needing breaks. Doesn't that kind of intelligence deserve respect in its own right?

Machines as Partners: Can we envision a future where AI isn't in competition with human intelligence, but acts as a powerful collaborator? Think of artists using AI to create things they wouldn't have imagined alone, or doctors using AI assistants to analyze complex medical data and find better treatments.

A New Philosophy: What we need is a whole new mindset when it comes to AI. One that doesn't start with the question: "Can it pass for human?" but instead asks, "What unique and amazing things can this technology do, and how can it make our world better?"

This way of thinking might lead to:

Rules for Respect: Creating guidelines ensuring machine intelligence is used for good and never to degrade humans as the "lesser" beings.

Celebrating the Strengths of Each: Encouraging AI development in areas where it complements humans, not simply copies us.

Redefining Progress: Shifting away from the idea that the only 'good' AI is the AI you can't tell apart from a person.

The uncanny valley might be uncomfortable, but it's also an opportunity. We can choose between a future where we either anxiously try to outdo our own creations, or one where both human ingenuity and machine intelligence are celebrated for their own unique brilliance. The path beyond the valley might be one where the goal isn't imitation, but amazing collaboration.

Conclusion: Living on the Edge of the Valley

Our journey through the uncanny valley has taken us to some strange and thought-provoking places. We've seen how the pursuit of hyperrealistic AI, while filled with technological marvels, forces us to confront deeply human questions:

The Nature of Connection: Can we feel true empathy for machines, however lifelike they may seem, and is there a danger in mistaking programmed affection for a genuine bond?

What Makes Us Real: In a world where AI might copy our creativity, our emotions, and even exceed us in some capabilities, how do we hold onto the importance of authentic human experience?

The Risk of Self-Replacement: If machines can one day do everything we do, and perhaps do it better, where does that leave the value we place on our own unique, yet imperfect, humanity?

The uncanny valley is a reminder that technological advancement is never just about what we can build, but also about the choices we make along the way. Do we endlessly pursue artificial imitation, leading to a future where the line between real and fake is impossible to find? Or, do we adopt a different mindset?

This new approach would champion AI designed to complement human skills, not mimic them. We'd create clear rules to ensure machines always serve humanity, never undermine it. Most importantly, it would be a future where we embrace the uncanny valley, accepting the differences between humans and our creations as a positive source of diversity and strength.

The relationship we forge with the increasingly intelligent machines we create is one of the defining choices of our time. This choice will determine not only what kind of technology fills our world, but also, it will shape the kind of humans we remain.

Let the uncanny valley serve as both a warning and an opportunity. It's a call to ongoing discussion, ethical development, and a commitment to a future where human brilliance and machine intelligence exist in harmony, each making the other even better.

About The Curious Philosopher

Welcome to The Curious Philosopher, your dedicated platform for diving deep into the world of philosophy. We are more than just a YouTube channel or a book publisher. We are a beacon of enlightenment, making complex philosophical concepts accessible and engaging for all.

Our YouTube channel is a rich repository of philosophy made simple. We take the profound and often complex ideas from the world of philosophy and break them down into digestible, easy-to-understand content. From the ancient wisdom of Socrates to the existentialist thoughts of Sartre, we cover a broad spectrum of philosophical schools and thoughts, making philosophy accessible to everyone, regardless of their background or prior knowledge.

As a book publisher, we take the same approach, transforming intricate philosophical theories into comprehensible narratives. Our books are not just collections of words, but vessels of wisdom that make philosophy approachable and relatable. We believe that philosophy should not be confined to academic circles, but should be available to all who seek to understand the world and their place in it.

At The Curious Philosopher, we believe in the power of curiosity and the pursuit of knowledge. We are here to stoke the fires of your curiosity, to guide you on your intellectual journey, and to help you navigate the fascinating world of philosophy.

If you are someone who is not afraid to question, to explore, and to learn, then you are in the right place. Join us on this journey of exploration, as we make philosophy easy to understand, one concept at a time.

Be sure to visit our Youtube channel at:

https://www.curiousphilosopher.com/youtube

You can also visit us on the web at

https://www.curiousphilosopher.com

Welcome to The Curious Philosopher. Stay curious. Stay enlightened.